SOUTH AFRICAN
WILD FLOWERS.

South African Wild Flowers

Illustrations by
A. BEATRICE HAZELL,
E. RITTER and D. BARCLAY.

Verses by
HERBERT TUCKER

Botanical Notes by
E. P. PHILLIPS, M.A., D.Sc., F.L.S.

INTRODUCTION.

A

THE Publishers do not pretend that this
souvenir of South African Wild Flowers
is in any way representative of the flora of
the country. Only a few specimens have
been selected, hap-hazard as it were, of some
of the most well-known and beautiful varieties in order to create an interest in the
subject, and to form a fitting souvenir for
friends and flower lovers.

They invited Mr. Herbert Tucker to write
some appropriate verses opposite each picture,
Dr. E. P. Phillips has also contributed some
valuable botanical notes for those who are
interested from a scientific point of view.
They are indebted to both these gentlemen
and to the artists for their valuable assistance
in producing this little work.

Not in Winter dies the year
In our Southern Hemisphere,
But in Summer's mid-career—

When the breezy uplands bear
Countless blossoms rich and rare
Laughing in the golden air?

Go then, floral missive, go?
Flash a gleam of Summer's glow
On the dwellers in the snow.

Would that we our friends could
     greet
With the very flowerets sweet,
Not this pictured counterfeit.

                              H T.

3

GLADIOLUS BREVIFOLIUS.
(Pijpjes.)

Painted ladies, gay and free—
Such in name and heart are we?
Wine-flushed Bacchanals of Spring.
On some grassy mountain steep
Joyous festival we keep,
And with the breezes sport and dance;
Flinging many a flaunting glance
To those who watch our revelling?

H. T.

4

ANTHOLYZA LUCIDOR.

(The Flowering Rush.)

'Mid the slopes' commingled green
My upstanding stem is seen.
Soldier-straight I hold my head,
Plumed with gay cockade of red.

H. T.

ERICA MAMMOSA.
(*Heath.*)

Of the brave legion of the heaths
    are we—

Each cohort by distinctive colours
    known.

Not least in fame, our scattered troops
    you see,

Staining the verdure with their
    sanguine tone.

DISA UNIFLORA.
(*The Red Disa.*)

When from some ferny ledge above
    the stream
That cleaves the dark ravine,
The wanderer sees your splendid
    blooms outlean,
As if self-poised on wings of crim-
    son gleam,
He gives glad homage to the richest
    gem
In our great Mountain's floral diadem?

H T

DISA GRANDIFLORA

RICHARDIA AFRICANA.
(*The Arum Lily.*)

O lilies! lilies! though like weeds ye
    grow

In every hollow, swamp and rain-
    soaked field,

Uplifting queenly blossoms, pure as
    snow,

Which broad leaves like a bodyguard
    beshield—

Yet still with ever new delight we
    trace

Your nobly simple sovereignty of
    grace.

H.T.

ROCHEA COCCINEA.
(The Red Crassula.)

A scarlet-coated sentinel I stand

'Mid the bare crags that crown the
mountain heights;

And watch the coming over sea and
land

Of golden days and starry-glittering
nights.

Love sends me now to distant lands
and strange,

Beyond my eagle-vision's furthest
range?

CRASSULA COCCINEA

A.HAZELL.

AMARYLLIS BELLADONNA.
(*The Belladonna or March Lily.*)

Hail, Belladonna! Well befitting name
By these bewitching goddess-flowers
    borne,
Whose white unvestured shapes, as if
    in shame,
Blush into rose like clouds of break-
    ing morn.

Ah! Well might Paris, gazing all
    intent
On your bright cluster with adoring
    eyes,
Hold back the apple in bewilderment,
Not knowing which fair queen de-
    serves the prize.

PROTEA CYNAROIDES.
(*The Giant Protea.*)

Uncounted flowers of every form and
    size

Lift their fair faces to our summer
    skies,

And many a starry shape on hill and
    vale is bright,

With massy disk and mien magestical,

The giant protea-monarch of them all—

Like a great sun expands upon the
    mountain height.

H. T.

LEUCOSPERMUM CONOCARPUM.
(*The Kreupel Boom.*)

Like quaint incurving candle flames
     that shine

Around some treasured shrine,

Fair golden stiles in guardian ring
     invest

The blossom's downy breast.

So with undying radiance in the heart—

Tho' seas and years may part—

Fond memory doth our earliest loves
     enfold,

   And the dear days of old.

H. T.

Leucospermum Conocarpum (Kreupel Hout)

NERINÉ SARNIENSIS.
(The Nerine.)

Though the Disa's scarlet pride
  Doth his Sovereignty declare,
Sweet Neriné! by his side
  Thou art throned, a Consort fair?
When thy petals, hued with flame,
  Flash their glint of golden sheen,
All the host of flowers acclaim
  Thee the floral pageant's Queen?

PROTEA LEPIDOCARPODENDRON.
(The Black-lipped Protea.)

Half hid in stiff-set leaves like some
strange nest,

Wherein young fledglings rest,

Peer forth the protea's quaintly fash-
ioned blooms.

For see! the hard-cased bracts, so
sere and brown,

Are lipped with sombre - hued and
feathery plumes,

And lined with softest down.

H. T.

PROTEA MELLIFERA.
(The Sugar Bush.)

When waning summer thickly studs
The sugar bush with long and taper-
    ing buds,
Gladly from day to day the sunbird
    sees
Their spires expand to rose-flushed
    chalices.
Then in their heart of down he dips
How eagerly his slender bill, and sips
The honeydew that fills each glowing
    cup
The leafy sprays for his delight hold
    up.

                             H. T.

PROTEA SPECIOSA.
(The Tawny-lipped Protea.)

Some treasure-trove methinks each
    spray has found—
  Some vase of price or quaint-
    wrought chalice rare;
So reverently the guardian leaves
    surround
  The glossy blooms their parent
    stems upbear.

H. T.

WATSONIA ANGUSTA.

(*Watsonias.*)

Not with blood in battle shed
Our uplifted spears are red,
Nor for any foe we wait in stream-
    fed valleys green;
But our banners bright we raise
In glad welcoming and praise
Of the sun - awakened Spring who
    comes to be our Queen.

H.T.

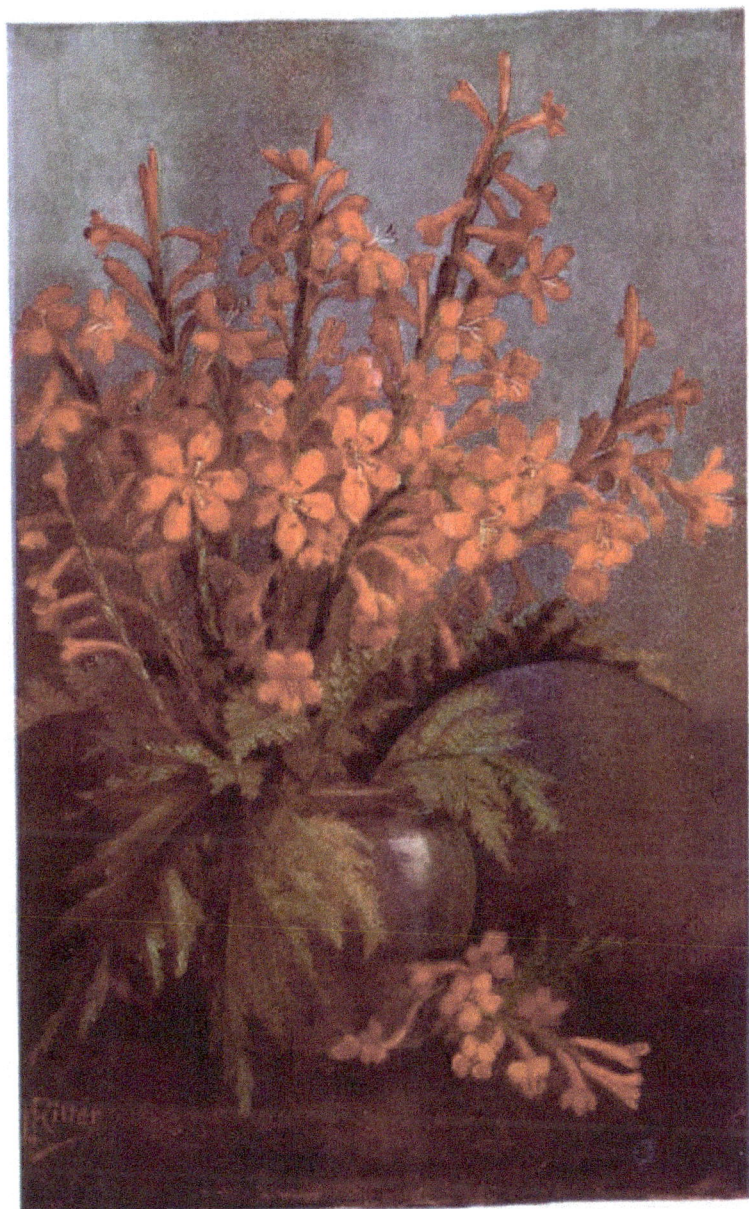

WATSONIA MERIANA.

(Watsonias.)

Loose-hung on ruby stems, your rosy
    blossoms swing

  On many a breezy height in the
      young summertime:

And to listening peaks the poet dreams
    they ring,

  Unheard of mortal ears, a glad melli-
     fluous chime?

H.T

HARVEYA CAPENSIS.
(*The White Harveya.*)

A name of scorn I bear—a parasite?

For from no earth-embedded stem I
    spring,

But to the roots of sturdier shrubs
    I cling,

And of their stolen sap my fibres drink,

And when with clustering blooms of
    purest white,

My life is crowned, their fragile petals
    shrink

From each ungentle touch as from a
    blow,

And like a bruise it stains their hue
    of snow.

H. T.

34

ERICA SPP.
(*Heath.*)

O myriad-blossomed heather clusters
    bright,
Gemming the veld with isles of red
    and white,
Like fairy lamps ye throng each taper-
    ing spray.
How beautiful it were, at fall of night,
If every tiny blossom leapt to light,
Illumining the murk with mystic ray?

                 H.T

GERBERA JAMESONI.
(*The Barberton Daisy.*)

Daisies many our land can show,
Yellow as gold or white as snow;
But wheresoever you may have been—
Answer me truly—have you seen
Any with hue so rich and rare
As I on my raying petals wear?
Barberton's pride! my ruddy star
Shines on the rugged slopes afar.

H. T.

DISA GRAMINIFOLIA.
(The Blue Disa)

O darling of the heights? When 'mid
    the wind-swept grasses,
Sways on its slender stem your
    wingéd hood of blue,
Where shall the wanderer find a blos-
    som that surpasses
    Your glad and lissome grace—your
    rich, translucent hue?

H. T.

BOTANICAL NOTES

by

E. P. PHILLIPS, M.A., D.Sc., F.L.S.

Gladiolus brevifolius, Jacq. The genus *Gladiolus* is largely represented in South Africa and forms a large proportion of the "bulbous" plants in the native flora. Different species are known by the names "Afrikanders," "Painted Ladies," and "Pijpies."

Antholyza lucidor, Linn. f. Over a dozen species of *Antholyza* are known from South Africa, some of them, viz., "The Flame" (*A merianella, Linn.*) and the "Red Africander" (*A. revoluta Burm.*) being of singular beauty. The species figured is common on stony slopes on the Cape Peninsula, and is also found in the Caledon and Uitenhage Districts.

Erica mammosa, Linn. A common heath at the Cape, being found in the Piquetberg, Paarl, Cape, Stellenbosch, Caledon, and Ceres Districts. The colour of the flower is variable ; varieties with purple-red, scarlet and white flowers are frequently met with. In habit it is an erect shrub up to 4 feet high. The species can always be recognised by the presence of four sacs or pouches at the base of the corolla.

Disa uniflora, Berg. Well termed "The Pride of Table Mountain." It may also be gathered on the mountains round Clanwilliam, Tulbagh, Worcester, Paarl, French Hoek, and Ceres. On the Cape Peninsula it is in flower from December to April, and inhabits the banks of mountain streams and wet krantzes. The specific name "grandiflora," with which the famous South African traveller and botanist, Carl Thuuberg, christened it, has unfortunately to be dropped in favour of the earlier name "uniflora." A variety with yellow flowers is found on the Cape Peninsula, but is rare. A handsome Cape butterfly (*Meneris tulbaghia*) sometimes visits this flower.

Richardia african1, Kunth. Known as the "Arum Lily" or "Pig Lily." The former name is rather inappropriate as our plant is neither an Arum nor a Lily. It belongs to a large family of plants (the Aroideæ), the majority of which are found in the tropics of both hemispheres. During the months of July to October, the Arum may be found in profusion on the Cape Peninsula, and usually inhabits damp shady situations. The snow-white trumpet-shaped structure is called the "spathe," the golden rod which it encloses is the "spadix." It is on the "spadix" that the real flowers are found, the female flowers near the base, the male above, so what is popularly called the "flower" is really a number of flowers massed on a cylindric rod and surrounded by a white "spathe."

Rochea coccinea, D.C. The "Red Crassula." It is frequently to be met with in flower on the Cape Peninsula in January and February, growing on rocks exposed to the fierce rays of the sun. *Rochea coccinea* is a member of a large family of plants known as the Crassulaceæ, which are very abundantly represented in the drier parts of South Africa. Most of them possess thick fleshy leaves which enable them to withstand periods of extreme drought without injury. The scarlet flowerheads are frequently visited by *Meneris tulbaghia*, which is attracted by red flowers.

Amaryllis Belladonna, Linn. The " Belladonna Lily " or
" March Lily," as it is popularly called, is not a true Lily, but
belongs to a closely related family (the Amaryllideæ). The
species figured is the only one we know of in the genus *Amaryllis*,
but the name *(Amaryllis)* is, though erroneously, in common
use among gardeners for species and varieties of the large
American genus *Hippeastrum.* On the Cape Peninsula it blooms
freely during the month of March (hence the name " March
Lily "). The flowers are strongly scented and are produced
after the leaves have withered.

Protea cynaroides, Linn. The " Giant Protea." This is
a very appropriate name as the " flower heads " are the largest
known among the Proteaceæ, and often measure over 6 inches
in diameter when fully expanded. It is to be found frequently
in various parts of the South-Western Region of the Cape
Province, and varies in habit from an almost stemless plant to
a bush over 6 feet high. On Table Mountain it flowers from
January to about April.

Leucospermum conocarpum, R. Br. The " Kreupel
Boom." A member of the protea family differing from the
genus *Protea* in the structure of the flowers. A bush in full
bloom is a sight never to be forgotten ; the bright canary-yellow
" heads " showing up in strong relief against the silvery leaves.
It is frequently seen in many parts of the Western Province,
but is largely destroyed for firewood. The flowers are visited
by " sun birds," viz,,*Promerops capensis* and *Anthobaphes violacea.*

Neriné sarniensis, Herb. Commonly cultivated in European
gardens under the name of " Guernsey Lily." Like *Amaryllis
belladonna* the Neriné is a member of the family Amaryllideæ.
Perhaps the most beautiful of the Cape Peninsula flowers ; the
petals which vary in colour from pink, orange, to a deep red
(and occasionally, though rarely, pure white), are covered with
a golden sheen. The Neriné is a bulbous plant, and may be
found in bloom soon after the first rains have fallen. At the
end of the flowering season the leaves are developed.

Protea lepidocarpodendron, Linn. The " Black-tipped
Protea." A common *Protea* on the Cape Peninsula, and is also
found in various parts of the Western Province. It grows to a
bush up to 8 feet high, and sometimes forms fairly dense
thickets on the flats and mountain slopes. The so-called
" flower " is made up of a number of flowers arising from a
convex receptacle and surrounded by numerous bracts (this
structure applies to all the species of *Protea*), the innermost of
which are densely covered with black hairs.

Protea mellifera, Thumb. The " Sugar Bush." When
seen in full bloom during June to August, the bush presents
a most gorgeous sight. The bracts vary in colour from pure
white, through shades of pink to a deep red, and are coated on
the outside with a sticky substance. The " flower " contains a
sweet watery liquor, which is collected by many Colonists, who
prepare from it by inspissation, a delicious syrup, which is
known as the *Syrupus Proteæ (Boschjes Stroop)*, and which is
stated to be of great use in cough and pulmonary affections.

45

Protea speciosa, Linn. A rather rare plant on the Cape Peninsula, but is distributed in many parts of the Western Province. This species does not attain the same height as the other species of *Protea* figured, but the singular beauty of its "heads" places it in the foremost rank of the "handsome" Proteas, over 80 species of which are known from South Africa.

Watsonia angusta, Ker. The genus *Watsonia* is represented by 16 species which, with one exception, are confined to South Africa. They are to be found in various habitats, some preferring damp marshy ground, others growing on the dry exposed mountain slopes. On the Cape Peninsula they may be found in flower from October to January.

Watsonia meriana, Mill. A variable species, of which the plant figured is a variety. It belongs to a large natural family, the Irideæ (Iris and Tulip family), the bulk of whose members are found in South Africa.

Harveya capensis, Hook. The "White Harveya." This plant is parasitic on the roots of other plants from which it draws its nourishment, and in consequence produces only very rudimentary leaves. It may be seen in bloom on Table Mountain during the months of November to February, but also occurs round Tulbagh, Stellenbosch, Caledon, Riversdale, and Humansdorp, and has even been found in Namaqualand. The delicate petals turn black if handled.

Erica spp. "Heath." The large genus *Erica* is represented in South Africa by over 450 species mostly massed in the South-Western Region of the Cape Province; they are absent from the Karroo, the Orange Free State, and the Transvaal, except on the high eastern mountains. The districts of Caledon and Riversdale are famous for many beautiful species which are found there, though unfortunately through indiscriminate plucking many of the more striking species have sadly diminished of late years.

Gerbera Jamesoni, Bolus. The "Barberton Daisy." This handsome plant, now largely cultivated, was first found in the Barberton district, Transvaal. It is a member of the family Compositæ (the Sunflower and Daisy family) and like the Proteas, the so-called "flower" is composed of numerous small flowers on a central disc, surrounded by a number of red rays.

Disa graminifolia, Ker. The "Blue Disa." This pretty orchid grows commonly in stony places among low bushes on the summit of Table Mountain, and flowers in February and March It is only one of the 47 species of *Disa* known from the Cape Peninsula. All the South African orchids, with the exception of a few found in the eastern forests, are terrestrial.

www.ingramcontent.com/pod-product-compliance
Lightning Source LLC
Chambersburg PA
CBHW051211090426
42740CB00022B/3469